< GET CODING! >

Get Coding with Hopscotch™

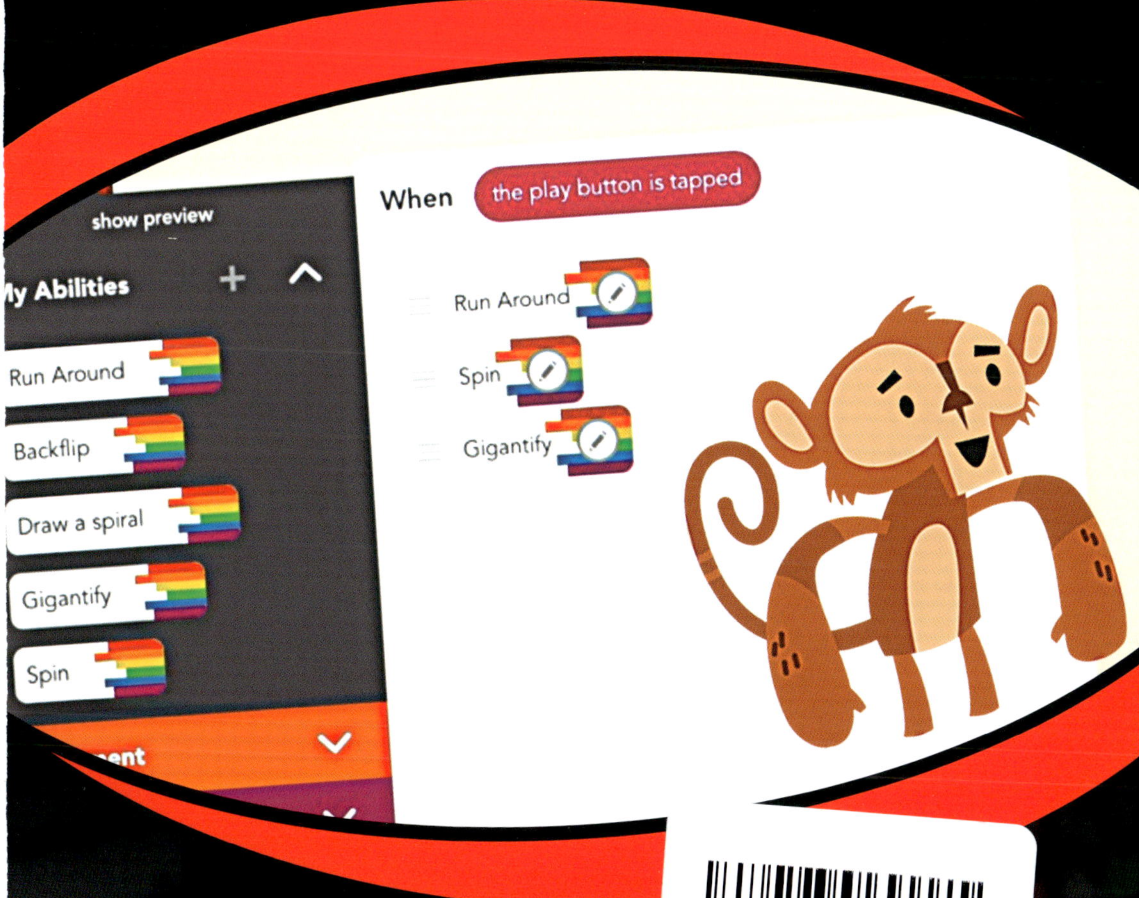

By Jenna

Gareth Stevens PUBLISHING

Please visit our website, www.garethstevens.com. For a free color catalog of all our high-quality books, call toll free 1-800-542-2595 or fax 1-877-542-2596.

Library of Congress Cataloging-in-Publication Data
Names: Vale, Jenna, author.
Title: Get coding with Hopscotch / Jenna Vale.
Description: Buffalo, New York : Gareth Stevens Publishing, [2024] | Series: Get coding! | Includes bibliographical references and index.
Identifiers: LCCN 2023004476 | ISBN 9781538288528 (library binding) | ISBN 9781538288511 (paperback) | ISBN 9781538288535 (ebook)
Subjects: LCSH: Hopscotch (Computer program language)–Juvenile literature. | Microcomputers–Programming–Juvenile literature.
Classification: LCC QA76.7.H57 V35 2024 | DDC 005.13–dc23/eng/20230207
LC record available at https://lccn.loc.gov/2023004476

Published in 2024 by
Gareth Stevens Publishing
2544 Clinton Street
Buffalo, NY 14224

Portions of this work were originally authored by Patricia Harris and published as *Understanding Coding with Hopscotch*. All new material in this edition authored by Jenna Vale.

Designer: Leslie Taylor
Editor: Megan Kellerman

Photo credits: Cover (illustrations) StonePictures/Shuttertstock.com, (Hopscotch imagery) courtesy of Hopscotch Technologies, Inc.; cover & series art (coding background) BEST-BACKGROUNDS/ Shutterstock.com; p. 5 Postmodern Studio/Shutterstock.com; p. 7 Pressmaster/Shutterstock.com; p. 9 karelnoppe/Shutterstock.com; pp. 11, 13, 15, 17, 19, 23, 25, 27 Hopscotch app screenshots; p. 17 Samuel Borges Photography/Shutterstock.com; p. 21 (2 screenshots) Permissions from/ Gethopscotch.com; p. cowardlion/Shutterstock.com.

Printed in the United States of America

CPSIA compliance information: Batch #CS24GS: For further information contact Gareth Stevens at 1-800-542-2595.

CONTENTS

Words in the glossary appear in **bold** the first time they are used in the text.

A Hop, Skip, and a Jump

When you think of coding, what comes to mind? Maybe you think of lots of math or endless lines of text, or words, on a computer screen. If that's what you think of, the prospect of coding can seem a little scary.

The creators of Hopscotch, a free coding **application**, or app, wanted to make coding not only easy but fun to learn. While many programming languages use text, Hopscotch uses colorful blocks to represent different commands. Users drag and drop the blocks to create **scripts** for artwork, games, and even websites. With Hopscotch, it's just a hop, skip, and a jump to understanding the basics of coding!

Hopscotch can be **downloaded** from the iOS store. It's only available for Apple products. Make sure you ask a parent or guardian before downloading any app, including Hopscotch.

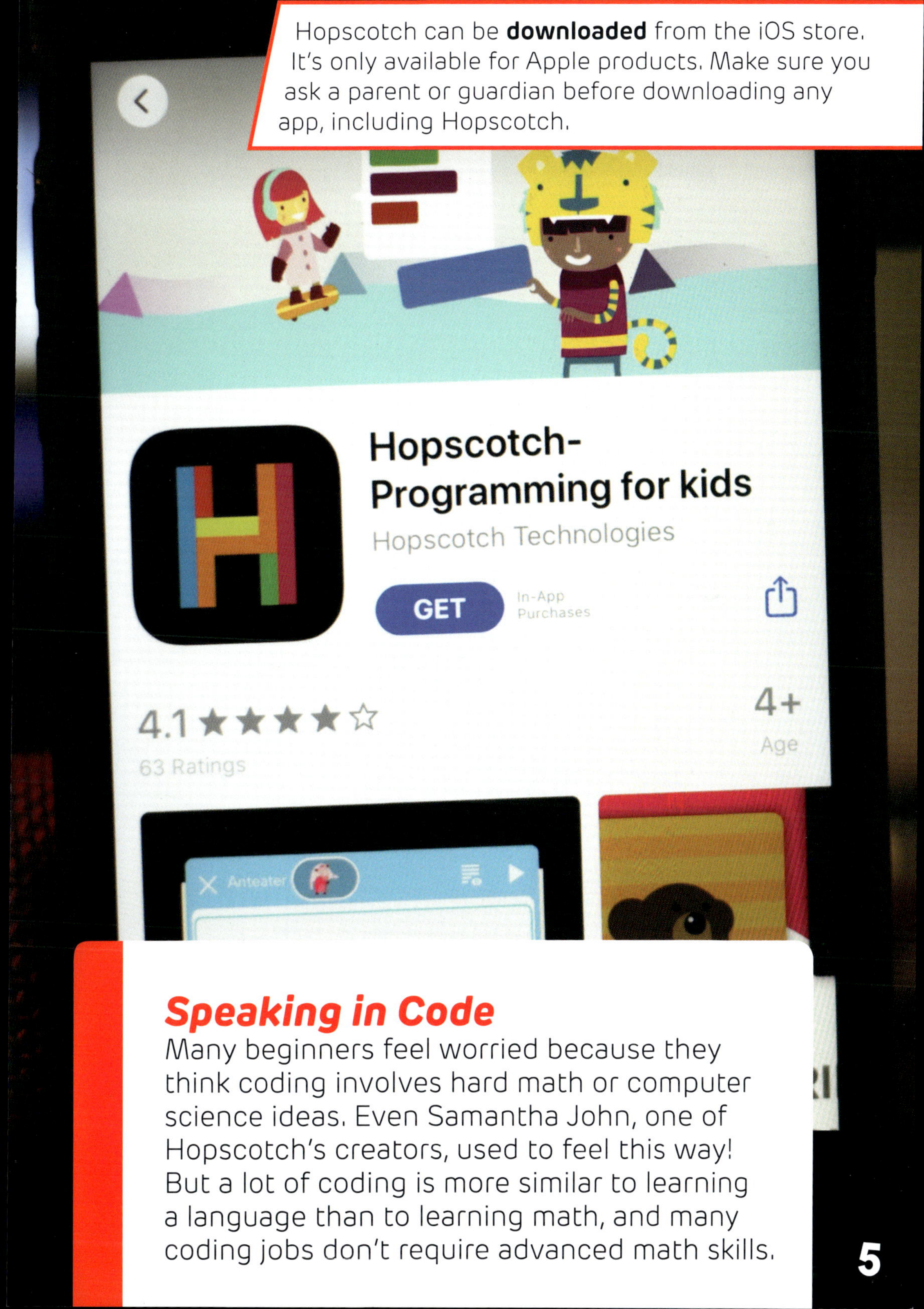

Speaking in Code

Many beginners feel worried because they think coding involves hard math or computer science ideas. Even Samantha John, one of Hopscotch's creators, used to feel this way! But a lot of coding is more similar to learning a language than to learning math, and many coding jobs don't require advanced math skills.

Hopscotch and Scratch

Hopscotch is similar to a programming language called Scratch. Both are free online apps that feature colorful blocks that users drag and drop to create scripts. The team that made Hopscotch wanted to make an app like Scratch that was even easier to use. They removed some kinds of blocks found in Scratch and added some new ones.

Programming languages such as Scratch and Hopscotch prove that anyone can code. They teach new coders, young and old, the basics of programming and prepare them for more advanced languages. The groups of blocks show users how to group commands while coding!

Hopscotch is a great app for coding your own games, making moving artwork, and creating your own apps.

Open-Source Software

Hopscotch and Scratch are open-source software. This means that users can freely share code they've written in those languages with each other. Software that is not open source is called proprietary, or closed-source, software. Proprietary software is owned by a company or individual and cannot be shared freely.

Learning the Rules

The first thing you need to know about programming in any language, including Hopscotch, is that coding is about following rules. That may sound a little like playing a game, and it can be just as fun.

Rule 1: Coders must plan what they want the computer to do.

Rule 2: Coders must use special words to have the computer take **input**, make choices, and take action.

Rule 3: Coders need to consider what tasks can be grouped together.

Rule 4: Coders need to apply **logic** using AND, OR, NOT, and other logic statements as key words.

Rule 5: Coders must explore the **environment** and understand how it works.

When you get the hang of basic coding rules, you can experiment in Hopscotch and have some fun!

Let's Get Logical

No matter what programming language you use, you also need to use logic when you code. The instructions you give to the computer should be as clear as you can make them. Clear instructions, or input, will help ensure the computer can perform the tasks you want it to do.

Choose Your Character

When you first log in to Hopscotch, get familiar with the app by going to the "Learn" section at the bottom of your screen. The **coding kits** you'll find there are fun and will help you understand some ideas you'll need.

To try coding an **animation** or game, pick a tutorial, or lesson, to follow. You'll begin each tutorial by choosing a character to add to the Editor, or workspace. Tap the "Add Code" button next to your character to begin coding. You can program your character to perform many exciting actions, such as jumping, dashing, and even growing or getting smaller!

Object and character models like the ones in Hopscotch are sometimes called sprites.

Planting a Seed

You can get started in Hopscotch without doing any tutorials, but if you want more tips beyond the Learn section, you can trade "seeds" for different tutorials to help your knowledge grow. Ask your parent or guardian for help if you need seeds to view a particular tutorial.

Hopscotch Blocks

There are six types of coding blocks in Hopscotch. Custom blocks are rainbow-colored blocks that allow users to select actions they've coded before. Movement blocks are red blocks that control how the character moves. Looks & Sounds blocks are green blocks that change what the character or background looks and sounds like.

Drawing blocks are purple blocks that allow users to draw lines and shade areas of the background. Variables blocks are yellow blocks that store information, or knowledge, and allow objects to share that information with each other. A variable is something that can be changed. Control Flow blocks are blue blocks that determine, or are the cause of, how often actions are repeated and how much time passes between movements.

To find the block you need, tap a category, or group, at the bottom of your screen, or swipe your way through the whole block **menu** to see all the options.

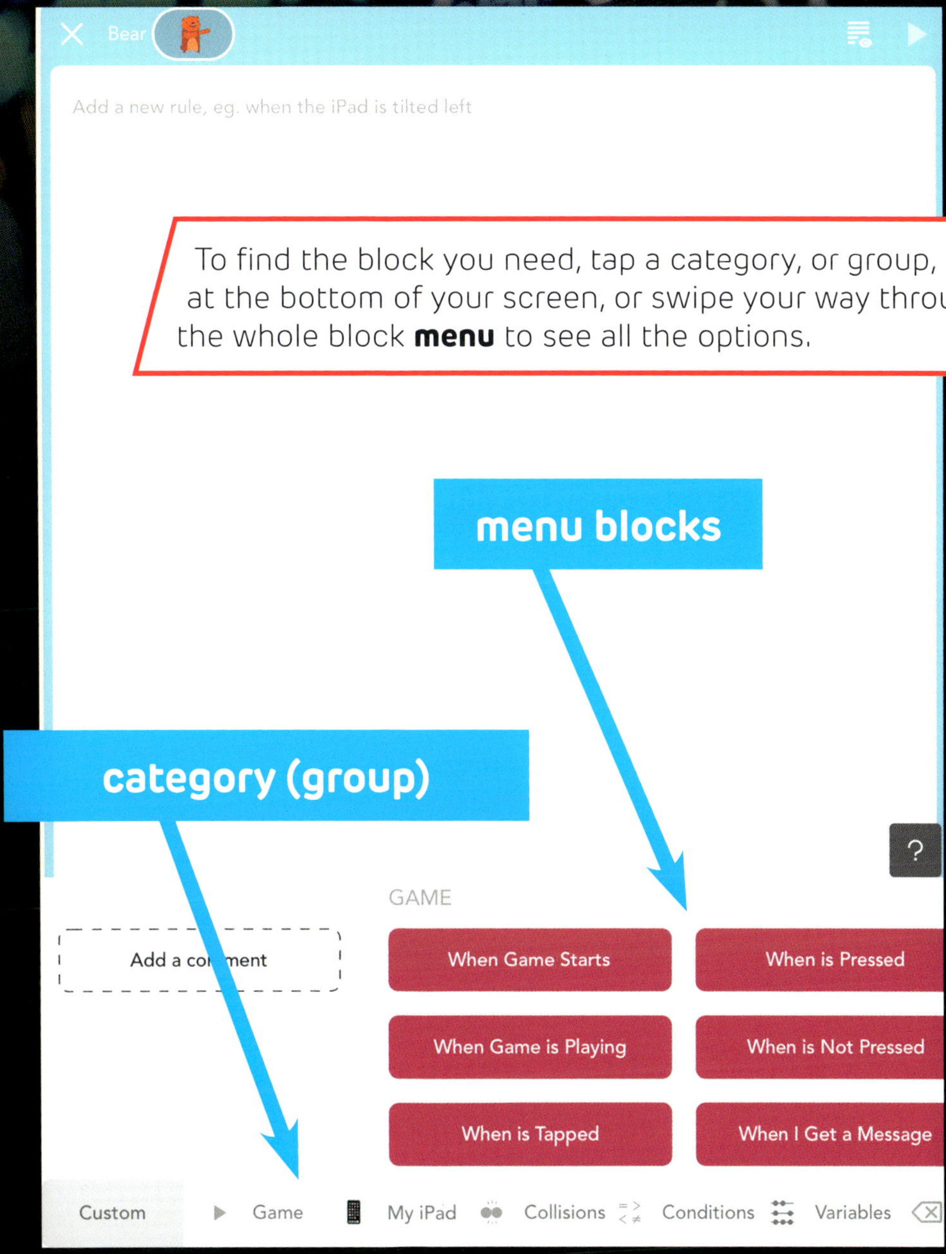

Sequencing

You'll write code by placing blocks in a sequence, or a group of things in a certain order. Sequencing is a fundamental concept for coding in any language. It involves putting your instructions for the computer in the order in which you want the actions to occur. In Hopscotch, the blocks are the instructions.

Time to Code

When you're ready to get started, make sure you're comfortable with the workspace. You'll see a picture of the screen for Hopscotch on the next page. In the bottom right corner of the screen, you'll see the "Create" button. When you tap Create, you'll begin a new project. The project will save automatically, or on its own, in the "Drafts" section as you make progress. Drafts are located in the top left of your screen.

The first step in your project will be to tap "Add Scene." You can build scenes for your project, and you can also copy, reorder, and rename scenes as you go.

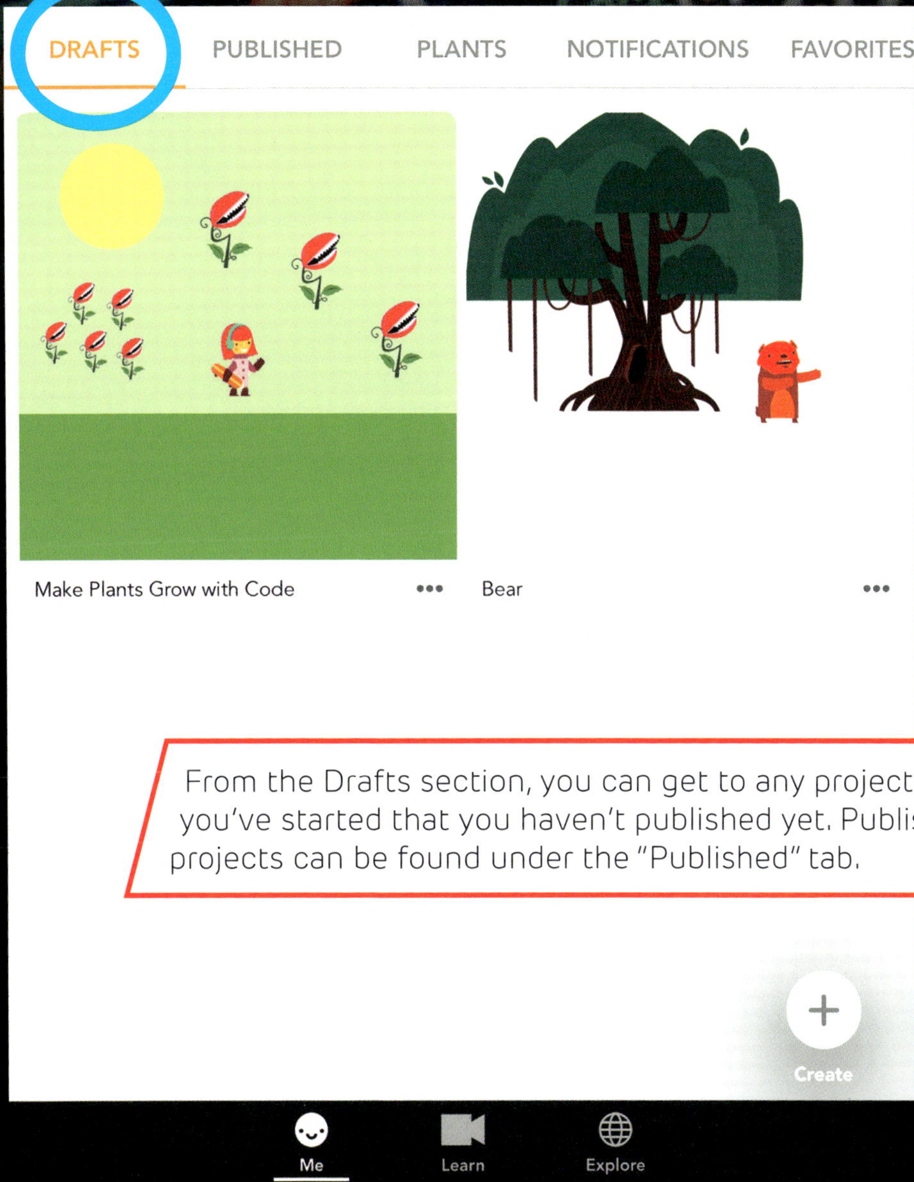

From the Drafts section, you can get to any project you've started that you haven't published yet. Published projects can be found under the "Published" tab.

And, Scene!

If you plan to make an interactive story or game, you'll probably want to add at least a few scenes to your project. The "Change Scene" variable block will help you code how and when a scene will change. You can label scenes with text or numbers—whatever works best for you!

Choosing What and When

In your first draft, decide what you want your character to do. Say you choose a bear and want to make it move around the screen. When you tap Add Code, you'll open the code menu and see the block menu at the bottom of the Editor.

The first block you'll use will be a magenta "When" block. A When block is used to **initiate** an action when a certain event occurs, such as when your game starts or when a user taps the screen. To make the bear move, first decide when you want the movement to start!

Figuring out the "whens" of your project is one of the most important steps of putting your code in order.

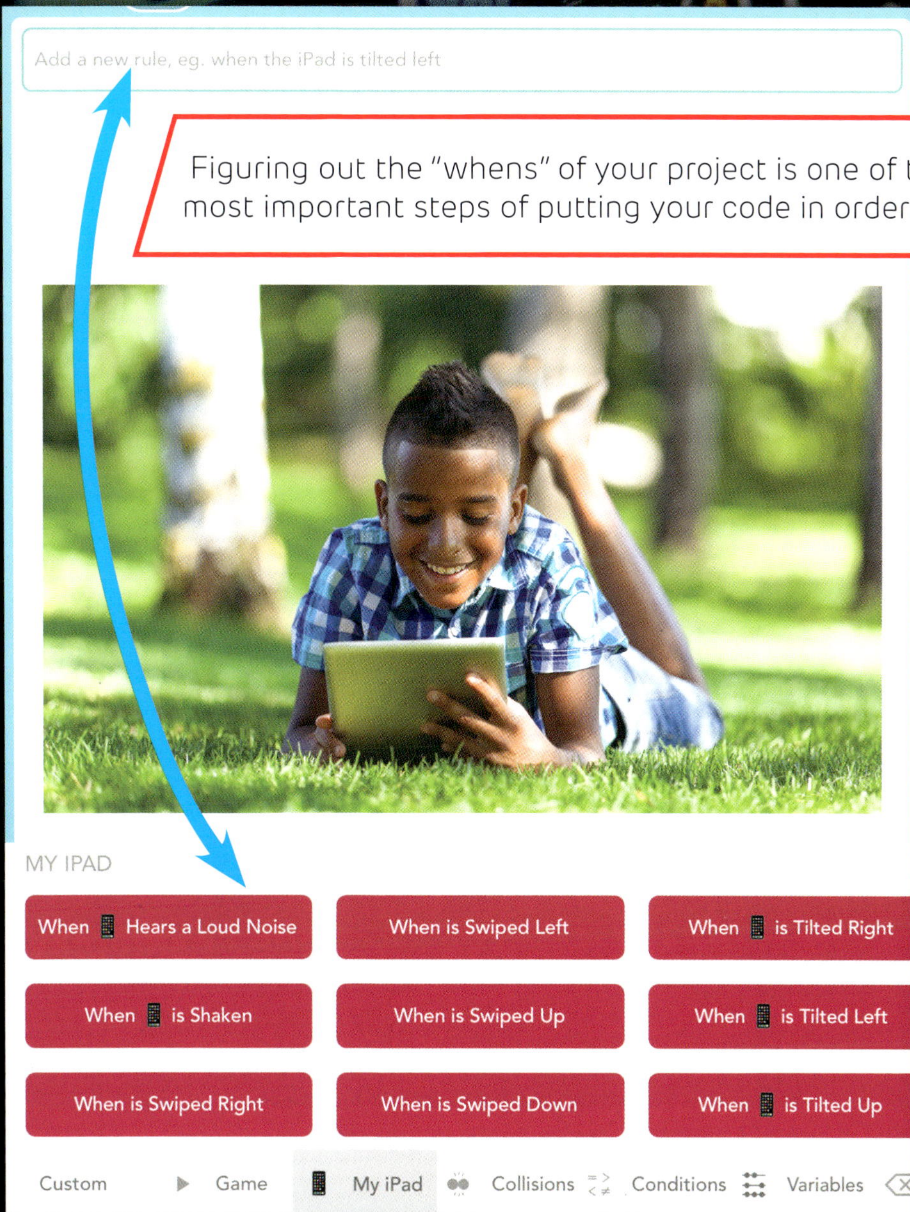

When X Happens, Then Y...

Hopscotch has an entire keyboard of "whens" for you to choose from. Check them out to see what's possible for your project. The three main groups of "when" you'll be able to choose from are when the user interacts with the device, when different kinds of **collisions** occur, and when certain conditions are met.

On the next page, you'll see one code scenario, or sequence, for this exercise. In this code, the bear begins to move right when the game starts. It will stop moving forward when it reaches the distance entered into the code, then it will "Change Pose" to dance in place and change to half its original size.

Next, you can try adding code for interaction between the user and the bear. You can make the bear disappear and reappear in a random spot whenever the user taps it—almost like it's running away from the user! You've just added **gameplay** to your first project.

Gameplay and Game Mechanics

Hopscotch is an excellent app for making your own games. When you make games, one of the things you add to the code are game mechanics. These are the rules of the game. The mechanics guide the player in how they can interact with the game and tell the game how to respond, or react, to player actions.

Bear with Me

This is the start of a game where the user chases the bear! What code would you add to create an ending to the game?

Code	What it does
When game starts	Sets the starting point for the program.
Move Forward 200	Moves the bear forward.
Change Pose	Makes the bear dance in place.
Shrink by percent 50	Makes the bear shrink, or get smaller, by 50 percent.
End	End of action that occurs "When game starts."
When Bear is tapped	Sets the new starting point of when the user taps the bear.
Random spot	When the user taps the bear, it will disappear and reappear in a random spot on the screen.
End	End of action that occurs "When bear is tapped."

Creating Loops

Deciding what happens and when is fundamental to writing your code, but adding every single step can make your code more cluttered than it needs to be. One of the ways coders avoid clutter is by creating loops. In Hopscotch, this means grouping actions together inside a blue "Repeat" block. Repeat blocks can be found in the Control Flow blocks menu.

You can use loops to repeat character actions, which can help you draw backgrounds for your project. The purple "Draw a Trail" block can be used to make a new background from scratch.

Import Your Own

If you have a background in mind but don't want to draw it from scratch, you can import, or add, your own image to Hopscotch from your device. In the same menu where you choose characters and objects, tap "Import" to upload, or bring in, an image as a background for your project.

You can even create two-color backgrounds. Create loops where two thin lines take turns being drawn back and forth repeatedly until the background is complete.

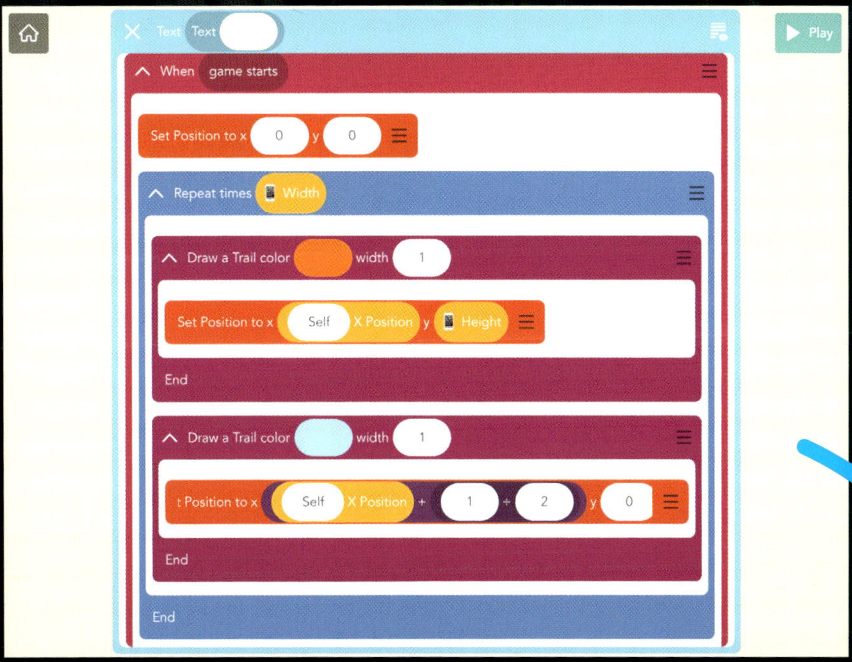

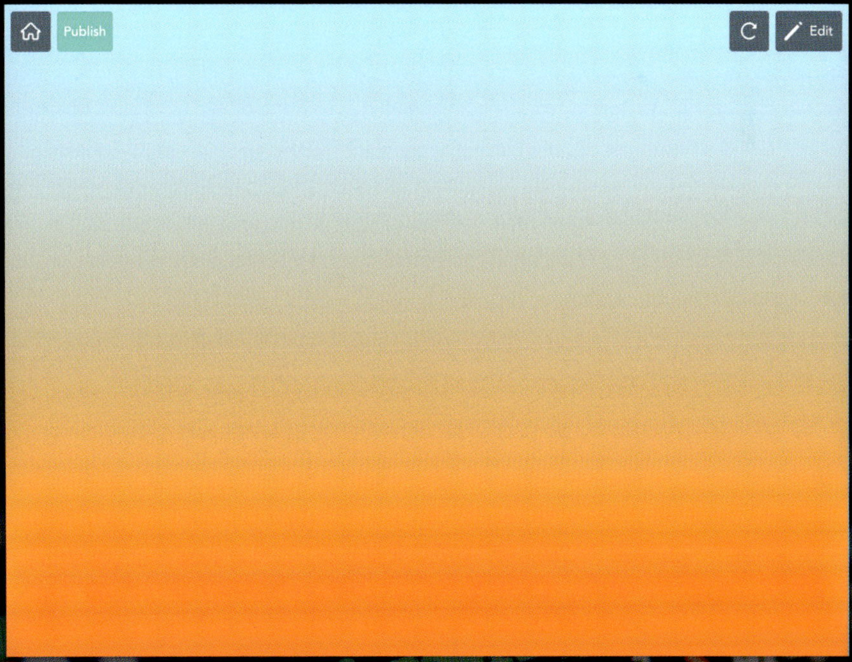

You can use the Draw a Trail block to practice making loops. See the example on the next page for a breakdown of the code. The goal of this exercise is to make the bear draw a trail in the shape of a square. You can code the bear to do this by putting a block for each movement in a sequence one by one, but then you'd end up with a very long line of code!

Think about which actions you can group and repeat instead. By creating a loop of the bear's actions, you create shorter, more manageable code.

Clone Some Help!

If you want to add many characters with the same behavior to your project, you can clone any character you've already coded with a blue "Create a Clone" block. Clones allow you to add more to your project without having to write as much code.

Hip to Be Square

Many blocks in Hopscotch can be used like containers. If you group actions within blocks, you can open or close the group by tapping the arrow to the left of the block's name.

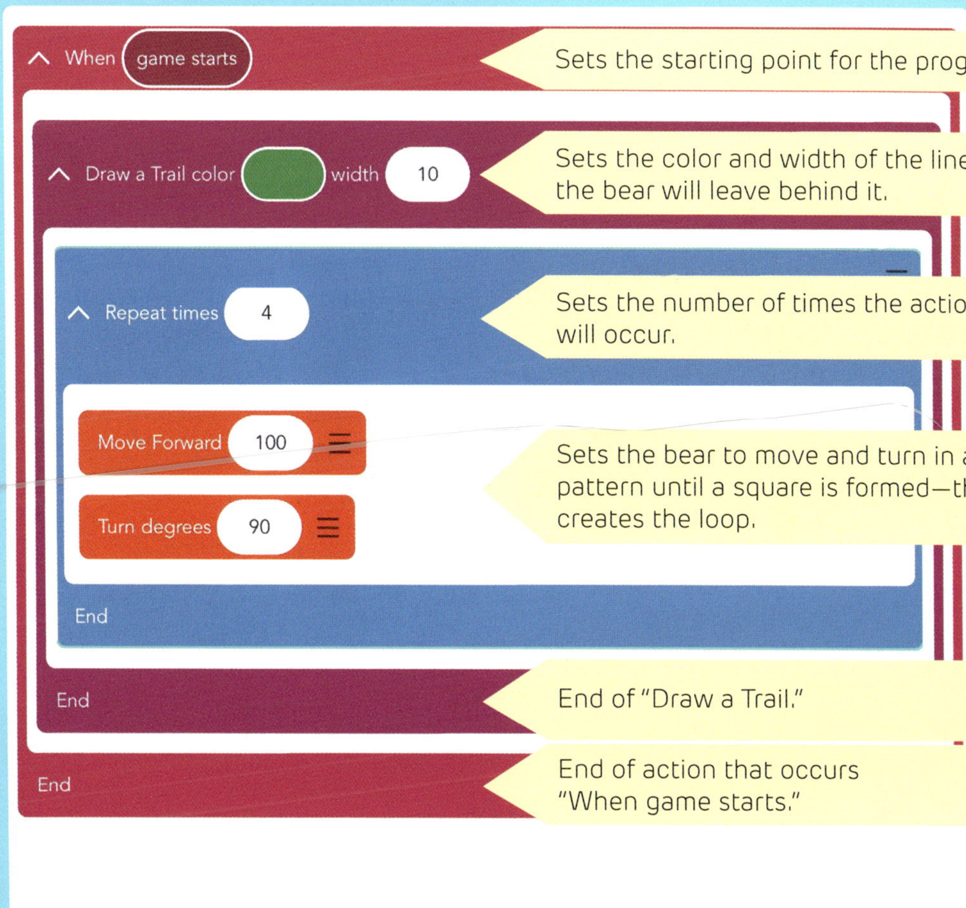

Sets the starting point for the program.

Sets the color and width of the line the bear will leave behind it.

Sets the number of times the action will occur.

Sets the bear to move and turn in a pattern until a square is formed—this creates the loop.

End of "Draw a Trail."

End of action that occurs "When game starts."

Block by Block

After you've had some practice, remember to follow these steps when you begin your first project:

1. Write out a simple plan.
I want a character to move across a scene.

2. Choose your character and import or plan your background.
I want a bear in a setting I've made from scratch.

3. Think about actions you can group together as a repeat.
I want the bear to move and change direction four times.

4. Think about any logic you would need to use.
I want my bear to walk AND turn.

5. Code!
Have fun with what you've learned and keep going!

Hopscotchers have made many great projects over the years that have pushed the limits of what can be created with Hopscotch. Find them on the Masterpieces Community Channel.

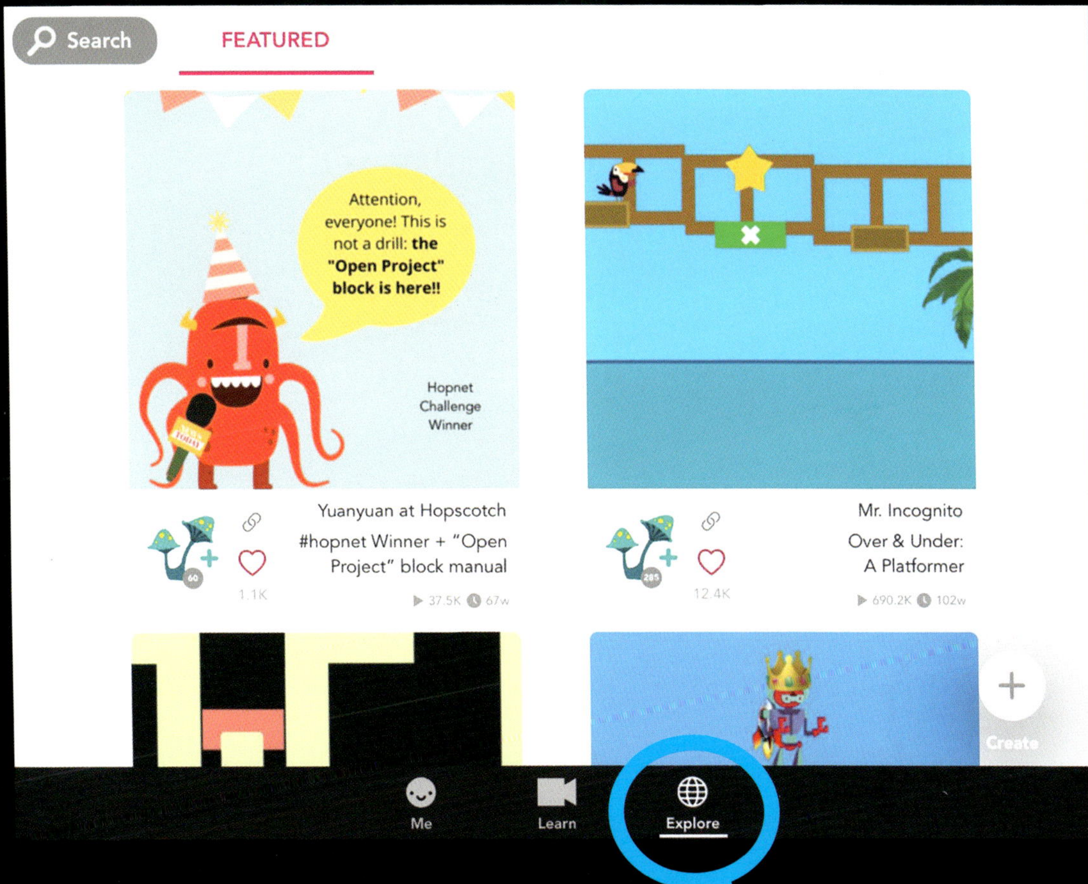

Learn From Others

Don't forget to check out tutorials and other projects to help you learn and create more exciting code. Hopscotch's Community Channels feature top games and challenges from the Hopscotch Team. They even allow you to follow other Hopscotchers. You can find Community Channels by tapping "Explore" at the bottom right of your screen.

A Group Effort

Publishing and sharing your work is one of the best parts of coding. Apart from the excitement of seeing what other Hopscotchers have made and adding your own projects to the mix, you can also seek help if you need to do any **debugging**.

If any part of your project isn't working the way you want it to, you can go through your code and long-press any block to find the "Help" option for more information. You can do this when viewing other users' code too. Seek out similar projects and view their code to help your own debugging process.

One of the best things about coding is sharing work, getting inspired, and encouraging others!

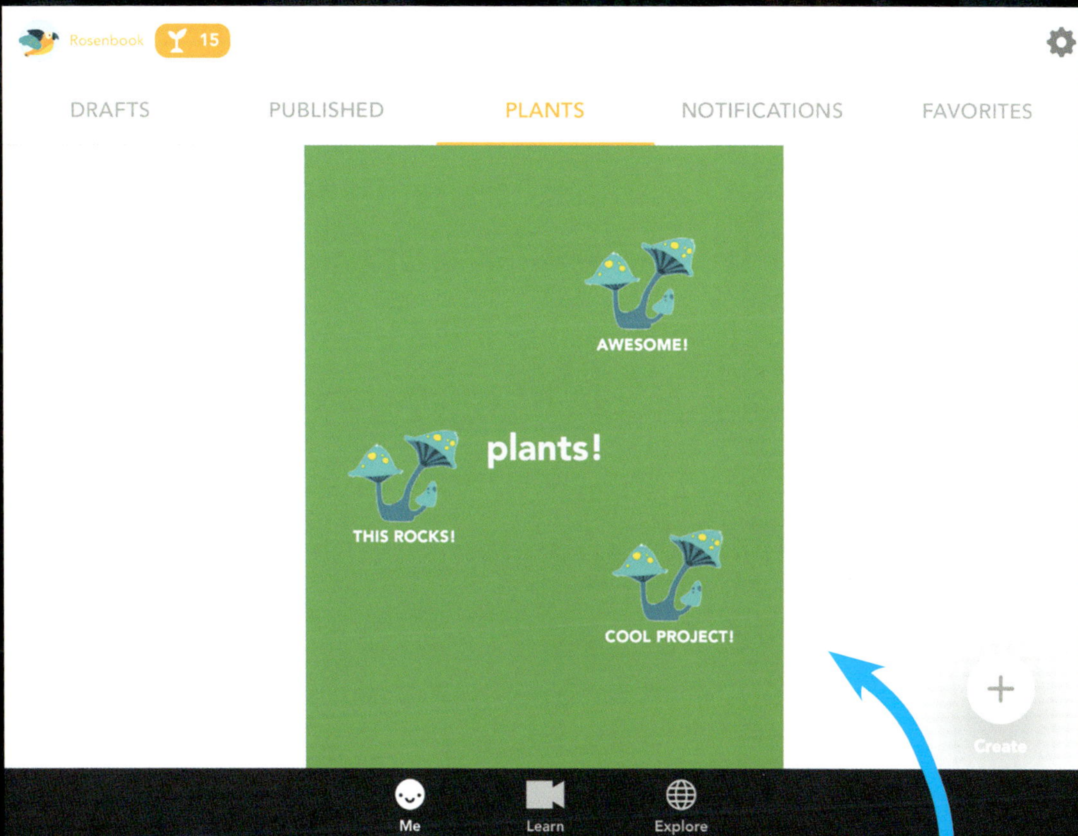

Help Others Grow Too

Remember the seeds from page 11? You can also use seeds to give "plants" to other users. If you see a great Hopscotch project, give a plant to the creator to thank them for sharing their work. You can find any plants others have sent you under the "Plants" tab.

Happy Hopscotching!

Hopscotch is one of the friendliest apps for anyone who wants to learn the basic building blocks of coding today. The style of Hopscotch's tools and user interface makes the app easy to navigate, or find your way through, and learn with—you may find that tapping, dragging and dropping, and taking out elements by moving them to the trash are all things you're already familiar with from other programs or apps.

Take your time learning and experimenting with code in Hopscotch. Explore the community or ask an adult for help when you need to. For now, you're ready to get started. Happy Hopscotching!

Learning to code is easier than most people think, especially with an app like Hopscotch.

Interaction Is Key

The term "user interface" describes the software elements a user interacts with. Hopscotch uses a graphical user interface, or GUI. When you use Hopscotch, pay attention to how you feel about manipulating, or using, tools and navigating the app. What you're feeling is called the user experience, which is another aspect of programming you're sure to run into again.

GLOSSARY

animation: Drawings or computer images that appear to move.

application: A software program, usually for a mobile device, that serves a specific purpose.

coding kit: An educational tool for learning programming concepts and applying them in a scenario.

collision: In gaming, the point at which an object runs into with another object or its environment.

debug: To find and solve issues, or "bugs," within lines of code.

download: To move or copy a file or program from one computer or device to another.

environment: The combination of computer hardware and software that allows a user to perform various tasks.

initiate: To begin.

input: Information that is entered into a computer.

gameplay: How a player or user interacts with a game.

logic: A proper or reasonable way of thinking about or understanding something.

menu: A list of things you can choose from.

script: A sequence of instructions carried out by a computer program.

FOR MORE INFORMATION

Books

McCue, Camille. *Getting Started with Coding: Get Creative with Code!* Hoboken, NJ: Wiley, 2020.

Scrivano, Álvaro. *Coding with Hopscotch*. Minneapolis, MN: Lerner Publications, 2019.

Smith, Grant. *Everything You Need to Ace Computer Science and Coding in One Big Fat Notebook: The Complete Middle School Study Guide*. New York, NY: Workman Publishing, 2020.

Websites

Code.org
www.code.org
An educational nonprofit focused on providing computer science tools for grades K–12.

GameDesigning.org
www.gamedesigning.org
Learn about game design from experienced designers, developers, and artists.

Hopscotch Forum
forum.gethopscotch.com
Hopscotch's online forum for collaborating, asking for help, and sharing tips for Hopscotch coding techniques and projects.

Publisher's note to educators and parents: *Our editors have carefully reviewed these websites to ensure that they are suitable for students. Many websites change frequently, however, and we cannot guarantee that a site's future contents will continue to meet our high standards of quality and educational value. Be advised that students should be closely supervised whenever they access the internet.*

INDEX